Lucas Soares da Silva
Neyla Cristiane R. de Oliveira
José de Ribamar de Sousa Rocha

Environmentalism at university: Biology students' perceptions

Lucas Soares da Silva
Neyla Cristiane R. de Oliveira
José de Ribamar de Sousa Rocha

Environmentalism at university: Biology students' perceptions

Environmental Knowledge

ScienciaScripts

Imprint

Any brand names and product names mentioned in this book are subject to trademark, brand or patent protection and are trademarks or registered trademarks of their respective holders. The use of brand names, product names, common names, trade names, product descriptions etc. even without a particular marking in this work is in no way to be construed to mean that such names may be regarded as unrestricted in respect of trademark and brand protection legislation and could thus be used by anyone.

Cover image: www.ingimage.com

This book is a translation from the original published under ISBN 978-613-9-66069-8.

Publisher:
Sciencia Scripts
is a trademark of
Dodo Books Indian Ocean Ltd. and OmniScriptum S.R.L publishing group

120 High Road, East Finchley, London, N2 9ED, United Kingdom
Str. Armeneasca 28/1, office 1, Chisinau MD-2012, Republic of Moldova, Europe
Printed at: see last page
ISBN: 978-620-7-98531-9

ACKNOWLEDGEMENTS

With respect, satisfaction and love, I thank you

To God, heavenly father, for always accompanying me in good times and bad, always guiding me and providing the possibilities to move forward in a constant process of strengthening and learning.

To the *Cosmos* for having made my existence possible, in the face of so many others, and the gradual understanding of such a complex reality.

To my parents, Ednaldo Carlos da Silva and Cristiane Soares de Oliveira Silva, for the concern and care they have always shown me, for loving me and accepting me with all my qualities and defects, for guiding me and never failing to show me my mistakes and successes.

To my family in general, sister Luciana Soares da Silva, for her permanent joy and relaxation, which gave me great happiness at various times; brothers Cicero Clésio Soares de Oliveira and Cleivan Soares de Oliveira, for the good times that gave me peace and joy; grandparents, for the wisdom and cosiness they gave me; uncles and cousins, for the various moments of relaxation and learning.

To all my friends and colleagues, who directly and indirectly contributed to my psychosocial development, opening up paths for me to understand the world around me.

To Professor and Supervisor José de Ribamar de Sousa Rocha, for his patience and devotion, for guiding me through this work, always providing me with paths to follow.

To all the teachers who have been part of my life, especially the professors at UFPI, for their example and dedication to the important act of teaching, even in the face of external or internal difficulties.

To all the other people who have been part of my life, for contributing directly or indirectly to shaping who I am today.

To me, for my persistence, hope and understanding!

SUMMARY

Various studies have affirmed the need to develop individual and collective awareness of the use of the environment, i.e. the whole formed by the space that encompasses all living beings, phenomena and physical elements, as well as the relationships between these factors. It is therefore necessary, a priori, to develop environmental knowledge, with a focus on preserving and conserving the environment. This is possible with the dissemination of environmental education, which should encompass all levels of education, with teacher training being its main building block and maintainer. The fact is that those teachers who have access to good training based on assimilating the most diverse aspects related to environmental knowledge would certainly make a more significant contribution to the provision of environmental education. In view of this, this study aims to demonstrate the state of environmental knowledge of undergraduate students in biological sciences. It also aims to verify: the existence or not of similarities in the learning of various environmental concepts; the degree of student satisfaction in relation to the environmental measures presented by UFPI and in relation to the course subjects focused on the environmental area; and the level of student interest in the search for the development of environmental knowledge. In order to ascertain the state of undergraduate students' knowledge of basic concepts in the environmental field, questionnaires were drawn up and administered to two undergraduate classes in biological sciences, one in the initial and the other in the final stages of the course. The questionnaire contained 20 questions on the subject of environmental knowledge, distributed equally, in quantitative terms, into four sub-themes: environmentalism, environmentalism at UFPI, selective waste collection and graduation. Each sub-theme comprised four multiple choice questions with three alternatives and one discursive question. The data obtained was tabulated and then analysed. With regard to environmentalism, by choosing the correct or most complete answers, most of the undergraduates demonstrated mastery of this subtheme. Considering environmentalism at UFPI, the undergraduates indicated the various measures that should be taken by the institution to preserve and maintain the environment. With regard to selective waste collection, the undergraduates had a good idea of the aspects of this topic, as the majority chose the correct alternatives. In the graduation sub-theme, the undergraduates demonstrated the importance of the course in shaping environmental knowledge, the need to improve the environmental approach in the subjects and their interest in this theme. It was also observed that the graduates from the concluding class have a greater mastery of this subject. The results obtained could serve as a basis for further research, with a view to streamlining the course curriculum.
Biological Sciences undergraduate course, regarding the disciplines in the environmental area, or even to take measures aimed at the adoption of environmental measures by the Federal University of Piauí (UFPI).

Keywords: Environmental education. Teacher training. Graduates.

SUMMARY

CHAPTER 1

INTRODUCTION

Environmental knowledge has never been as well established as it is today, not least because it is necessary to explain the various positive and, above all, negative aspects of human interaction with the environment around us. This knowledge has been built up procedurally, undergoing various changes that were generally linked to the situations and needs of each era of human society, in other words, according to a set of cultural knowledge and values (FLORIANI, 2008).

For example, human civilisation in its early days was directly related to the natural environment, living with and cultivating the environment around it, or rather, living in balance with the environment. Over time, the various transformations, whether cultural or technological, ended up changing man's relationship with his environment, making him linked to the need for ever greater production and consumption. Production and consumption, in turn, were (and are) only possible if there is an environment, which provides all the raw materials. So the focus of interest shifted from the environment necessary for survival to the environment geared solely towards production, without any or minimal concern for the results generated. It is in this changed focus that environmental knowledge has developed, but it has not been portrayed or taken as the most important, especially with regard to one of its precepts: preservation or sustainability. The lack of importance given to environmental knowledge was mainly due to a cultural aspect based on classical geography (GUIMARÃES, 1995; CUNHA; GUERRA, 2007).

Today's society reflects almost the same situation as a result of the focus on inconsequential production and consumption, i.e. where environmental knowledge is vast but little publicised or used.

Reality shows that environmental knowledge is necessary not only in the process of developing individual and collective awareness, but also in the process of society's socio-economic development. Some of the major problems related to environmental knowledge are: the way it is transmitted, the importance attributed to it and the way it is perceived (environmental perception). Generally, in a world geared towards unbridled consumerism, the importance given to knowledge of the various associations and definitions linked to environmental knowledge, as mentioned above, is not so evident or only exists in a masked form, in order to circumvent demands caused by the social sphere itself (CUNHA; GUERRA, 2007; FLORIANI, 2008).

As for the transmission of environmental knowledge, the problem relates to its absence or distorted occurrence based on a traditionalist character, with education being the best way of bringing it to light (CUNHA; GUERRA, 2007). Environmental education, in particular, first appeared in an integrated way in legislation with Law 6.938 of 1981, which instituted the National Environmental Policy. This law was later adopted by the Federal Constitution of 1988, which incorporated the concept of sustainable development dedicated to the environment. Based on the fact that the Constitution states that everyone has the right to an ecologically balanced environment, which is an asset for the common use of the people and essential to a healthy quality of life, it is necessary for the public authorities, together with the people, to defend and preserve this environment for current and future generations (RAMOS; COSTA, 1992). But in order for this right to be guaranteed, the government must, among other things, promote environmental education at all levels of education and raise public awareness of environmental preservation. Unfortunately, this is not the case, and environmental education, as well as the transmission of environmental knowledge, is only evident or restricted to basic education, with almost no manifestation in the community and/or higher education institutions, which in fact, in the latter, is also a major problem, especially when linked to the training of professionals. Basically, environmental education becomes important when it emphasises the need for integration between human beings and the environment, an integration that is harmonious and balanced, and which makes it possible to transform the current environmental scenario (GUIMARÃES, 1995).

Realising that today's reality shows a world inundated with problems, including environmental ones, and where these problems are present with increasing frequency, it is necessary to implement measures aimed at eliminating them completely. Among the measures: reducing and cleaning up pollution, reducing consumption of fossil fuels, conservation and sustainable use, protection of representative or unique ecosystems, among others; the one that is most urgent or necessary is the development of environmental awareness in people. In view of this, the development of this awareness must take place in the educational environment, at the various levels of education (GUIMARÃES, 1995).

Studies carried out on the Biological Sciences degree course at the Federal University of Piauí (UFPI), such as the study by Santos (2013), which deals with monitored visits as a technique for teaching environmental education, demonstrate the efficiency of contact with the reality of the environment and the very approach provided by education for the development of environmental knowledge and environmental awareness.

The fact that there is a need to know about the development of environmental knowledge in teacher training in Biological Sciences; the need to have concrete facts about the level of knowledge that students have in the environmental area; and the need for discussions about improving the development of environmental disciplines within the course already mentioned, make it necessary for this work to be developed with the proposed theme fitting into the context of research related to the environmental knowledge of biologists in training at UFPL.

In view of the above, the general aim of the following study was to check the state of undergraduate Biological Sciences students' knowledge of basic and more complex concepts in the environmental field, at UFPI's "Ministro Petrônio Portella" Campus in Teresina-PI, covering two undergraduate classes: a beginner and a graduate. The term "beginner" is taken to mean students who are in their initial periods of the course in question, and the term "graduate" is taken to mean students who are in their final periods. The specific objectives were: to check for difficulties in learning the various environmental concepts; to identify the level of interest of undergraduates on the Biological Sciences degree course in topics related to the environmental area; and to check student satisfaction in relation to the subjects on the Biological Sciences degree course that focus on the environmental area.

CHAPTER 2

THEORETICAL BACKGROUND

2.1   The importance of preserving and conserving the environment

The environment as a whole is closely related to the development and continuity of the dynamics of life on the planet. Human beings also take part in this dynamic process as they build themselves and society, thus emphasising the importance of their relationship with the environment. Cunha and Guerra (2007) emphasise this by mentioning that the man/nature dialectic is at the basis of the process of development and transformation of human societies.

The relationship between these two agents is of the utmost importance and has changed over time. Human beings have gone from being part of nature, participating in it and preserving it, to developing an individual consciousness, in which they become part of the whole and disintegrated from the equilibrium relationships in the environment, as mentioned by Guimarães (1995).

With this transformation, human beings changed their concepts and began to believe in a nature that hinders the development of society as a whole. At this time of change, space was opened up for the emergence of environmental problems, imbalances in the environment, which also hurt the agent who caused them (GUIMARÃES, 1995; CUNHA; GUERRA, 2007).

Despite the prolongation of the problems caused by the change in human awareness of their existence and relationship with the environment, recently we have seen the emergence of environmental awareness with the elucidation of society itself around the limitation of the resources offered by nature and the disastrous end that this can generate for their own existence and that of other beings who share the same planet (CUNHA; GUERRA, 2007).

In view of this, one of the principles in the development of environmental awareness lies in the act of preserving and conserving the environment. However, the difference between the two terms must be understood. The term environmental preservation, linked directly to the preservationist strategy, refers to the integral protection of nature or the environment, without any use of its resources Terborgh (1999 *apud* CUNHA; GUERRA, 2007) emphasises this by mentioning that there is an intrinsic value in nature that should be preserved for its own sake and not as a reserve of natural resources for the use of human beings. Conservation, linked to the conservationist strategy, deals with the protection of nature through the sustainable use of natural resources, i.e. the productive use of these resources goes hand in hand with their conservation, with a view to benefiting present and future generations (HALL, 2000 *apud* CUNHA; GUERRA, 2007).

Although the definitions of preservation and conservation differ, the actual practice of both is of the utmost importance to the environment. There needs to be a transformation from a society whose development is based on the unlimited exploitation of natural resources, to one whose development is integrated with the well-being of the environment. Protecting the environment from excessive exploitation is one of the best ways to maintain its dynamic balance. And by establishing the continuity of this balance, we can ensure the continuity of the beings that take part in this process, as well as the relationships that occur between them. Conservation of the environment and the development of collective awareness allows the present to be maintained and guarantees the possibility of a good future (CUNHA; GUERRA, 2007; GUIMARÃES, 2010).

However, all this practice can only take place if the individual has mastery of environmental knowledge, not only in terms of the various definitions, but also the relationships that are developed between them. It is in this context that environmental education comes in. According to Guimarães (2010), it focuses on the dynamic balance of the environment, in which life is perceived in its full sense of interdependence of all the elements of nature. Still according to Guimarães (2010), environmental education has the important role of fostering the perception of the necessary integration of human beings with the environment.

2.2 Environmental education

The dissemination of environmental knowledge is of the utmost importance for raising people's individual and collective awareness. This dissemination, in turn, is best achieved through environmental education, which should cover the various levels of education, especially those related to the training of educators.

According to Guimarães (2010, p.9), environmental education, in its scope,

> [...] presents a new dimension to be incorporated into the educational process, bringing a whole new discussion on environmental issues, and the consequent transformations of knowledge, values and attitudes in the face of a new reality to be built.

The aforementioned author basically states in his studies that one of the main reasons for the need for environmental education is due to the fact that humanity has distanced itself from nature, from the environment, over the course of time, in order to turn to the consumerism of resources, capital and goods. Still according to the same author, he believes that in order for there to be truly environmentally sustainable consumption, it is necessary to change consumerist values and not just the right attitudes. It could even be argued that this change in values can be directly generated by environmental education.

Focussing on higher education courses to train educators, the way in which this environmental education is applied is directly related to the level of environmental knowledge of future educators, which in fact is extremely important in the educational environment. And this is in fact directly related to the way in which universities work with this context in their training courses, especially those for educators. According to Teixeira et al. (1998 *apud* LOUREIRO; LAYRARGUES;CASTRO, 2006), as well as disseminating knowledge and preparing professionals, universities must maintain an atmosphere of knowledge, preparing people for the conservation of living knowledge and the process of their own awareness and progression.

According to Loureiro, Layrargues and Castro (2006), the training of environmental educators (influenced by environmental education in universities) is extremely necessary and urgent, mainly due to the growth of environmental awareness as a result of environmental problems and the importance that educational systems have in the search for indirect solutions to such problems. The guidelines of the Tbilisi Conference (1977), quoted by Loureiro, Layrargues and Castro (2006), state that universities should include environmental education in their teacher training courses, as well as supporting it, and that they should facilitate appropriate environmental training that is accessible to all future teachers.

Although several changes have already taken place, mainly leading to different reformulations in teacher training curricula, dissatisfaction is still very evident when it comes to the development of environmental knowledge. Loureiro, Layrargues and Castro (2006) state that the training of environmental educators implies a methodological, conceptual and curricular reformulation or even a new type of teacher, where the educator must work with knowledge in a way that employs the logic that results from the mutual interaction between various factors such as: the subject, the object of knowledge, the affective dimension, the vision of complexity and the contextualisation of environmental problems.

The need for changes and additions to the curricula themselves, through universities, are present and necessary in the training of environmental educators, who are loaded with the most diverse knowledge in the environmental field. But before such changes can take place, concrete data is needed on the influence of environmental education on the curricula of teacher training courses. To this end, various projects or even research linked to the training of teachers themselves can be developed to provide relevant data for the process as a whole.

CHAPTER 3

MATERIAL AND METHODS

3.1 Place and group studied

The study was carried out at the Federal University of Piauí (UFPI), on the "Ministro Petrônio Portella" Campus, in Teresina, Piauí, from April to June 2014. The groups studied were represented by two classes of undergraduates on the Biological Sciences degree course. The first class (Class A) was made up of beginners (3° periods), from the initial periods of the course, and the other class (Class B) was made up of graduates, from the final periods.

3.2 Data collection

Firstly, in order to carry out this work, we carried out a bibliographical survey related to environmental education, mainly aimed at teacher training, as well as the various concepts associated with environmental issues. It was noted that environmental education is based on teacher training. With this in mind, it was decided to draw up a questionnaire for data collection containing 20 questions organised according to various concepts associated with environmental issues. It was divided into four sub-themes: environmentalism, environmentalism at UFPI, selective waste collection and graduation. The questions were distributed equally in quantitative terms for these sub-themes, each with four multiple choice questions with three alternatives, and one discursive question. Before the questionnaire was administered, the interviewees were presented with a consent form, which was necessary for the questionnaire to be administered.

Two models of questionnaires were drawn up, entitled "A" and "B". The first was aimed at undergraduates in biological sciences who were in the initial periods of the course, and the second at undergraduates who were in the final periods. The only difference between the two questionnaires was in the first and fourth questions of the graduation sub-theme.

The questionnaires were administered to the two undergraduate classes, in classrooms, with the authorisation of the teachers who accompanied them.

3.3 Analysing the data collected

Once the questionnaires had been administered, the data was tabulated and transferred to spreadsheets using the Microsoft® Excel® 2010 programme. The data from the multiple-choice objective questions and the discursive questions were organised in quantitative and percentage terms, generating tables and graphs. In addition, the similarity between the alternatives chosen by the students taking part in the survey was also calculated in percentage terms. This similarity was calculated using the Jaccard and Sorensen similarity indices, which were adapted for the research.

3.3.1 Jaccard's similarity index or coefficient

It considers the relationship between the number of common answers and the total number of answers found when comparing the two classes (MUELLER-DOMBOIS; ELLENBERG,1974).

It is given by the formula:

$$Cj = \frac{a}{a + b + c}$$

3.3.2 Sorensen's similarity index or coefficient

It is based on the relationship between the double number of common answers and the total number of answers found in the two classes (MATTEUCCI; COLMA, 1982).

Generally, this index has higher values than Jaccard's, because the answers that are common between the classes when compared are given greater weight than the answers that are exclusive to one class (SILVESTRE, 2009).

It is given by the formula:

$$Cs = \frac{2a}{2a + b + c}$$

In both formulae, "a" corresponds to the number of answers common to both classes; "b" corresponds to the number of answers exclusive to class A; and "c" corresponds to the number of answers exclusive to class B. By analysing and discussing the data, the environmental knowledge of undergraduate students in Biological Sciences at UFPI and its characteristics were verified.

CHAPTER 4

RESULTS AND DISCUSSION

The results obtained from analysing the four sub-themes associated with the environmental theme were as follows:

4.1 Environmentalism

The first sub-theme on which we worked to diagnose environmental knowledge was environmentalism. Five different questions were organised, the first of which is subjective and the others are objective multiple choice questions.

In question 1, an open question, the answers were grouped together to form four specific groups (Table 1).

Table 1. Definitions of the environment according to undergraduates at the beginning and end of _the Biological Sciences degree course at IIFPI.

Question 1 - What concept would you give to the environment?		Class A		Class B		similarity index	
	Answers	N°	%	N°	%	Jaccard	Sorensen
1	Space where living beings live and relate to each other.	8	38,10	10	45,45	80	88,89
2	The set of factors that make life possible, such as living beings themselves and the environments they inhabit.	7	33,33	11	50	63,64	77,78
3	Space where human beings live or the whole of nature.	2	9,52	1	4,55	50	66,67
4	No answer	4	19,05	0	0	0	0
	Total	21	100	22	100		

Source: Direct research.

In the first group, which has the information that the environment is the place where living beings live and relate to each other, the percentages were 38.10% for class A and 45.45% for class B. In this option, the similarity was 80% and 88.89%, respectively for Jaccard and Sorensen, indicating that the students in the two classes chose this definition in a similar way.

The second group of answers, in which the environment is described as

as the set of factors that make life possible, such as the living beings themselves and the environments they inhabit; obtained a 33.33 per cent choice for class A and a 50 per cent choice for class B. The Jaccard and Sorensen similarity indices for this option were 63.64 per cent and 77.78 per cent. It can be seen that there was a drop in the similarity between the answers, which is mainly due to the fact that more students in class B chose the second alternative.

The third group covers the definition that the environment is the space where human beings live or the whole of nature. In this group, the percentage of choices was 9.52 per cent in class A and 4.55 per cent in class B. The similarity indices were even lower compared to the second group, with 50% and 66.67% for Jaccard and Sorensen. This difference is due to the fact that fewer students in class B opted for this alternative.

The fourth group comprises only those students who didn't answer the question, for various and variable reasons such as lack of knowledge or interest. Only the students in class A are included in this group, totalling 19.05%.

Among the objective questions, question 2 asks students what concept they would attribute to the environment; question 3 asks which statement would best characterise biodiversity in terms of the changes that could occur; question 4 asks what the best definition of environmental impacts would be; and question 5 asks what the best definition of global warming currently is (Table 2).
Question 2 proposed three different alternatives. Analysing Table 2, it can be seen that of the 21 students in class A who took part in the survey, approximately 52.38% gave the diversity of living beings as the best definition for biodiversity; while of the 22 students in class B, 36.36% chose this alternative. In class A, approximately 28.57% chose the association of various hierarchical components: ecosystem, community, species, populations and genes in a defined area; and in class B, 40.91% chose the same alternative. Of the students in the beginners' class, 19.05% attributed the totality of genes, species and ecosystems in a region; while of the students in the finalists' class, 22.73% chose this definition.

Still related to this question, according to Article 2 of the Convention on Biological Diversity (BRAZIL, 2002), the term biodiversity can be understood as the variability of living organisms from all sources, encompassing terrestrial, marine and other aquatic ecosystems, including their ecological complexes; as well as diversity within species, between species and of ecosystems. In view of this, when discussing question 2, there is talk of more complete or less complete concepts.

Table 2. Environmentalism according to undergraduates from the initial and final periods of the Biological Sciences degree course at UFPI.

Question 2 - According to your knowledge, the term biodiversity can best be described as:	Class A		Class B		Similarity index	
	N°	%	N°	%	Jaccard	Sorensen
A) Diversity of living beings.	11	52,38	8	36,36	72,73	84,21
B) Association of various hierarchical components: ecosystem, community, species, populations and genes in a defined area.	6	28,57	9	40,91	66,67	80
C) The totality of genes, species and ecosystems in a region.	4	19,05	5	22,73	80	88,89
Total	21	100	22	100		
Question 3 - In relation to the changes that could occur to this biodiversity, what can best be said?						
A) Biodiversity can be seen as static.	1	4,76	0	0	0	0
B) Biodiversity is immersed in a dynamic of changes that occur all the time.	17	80,95	19	86,36	89,47	94,44
C) It can be said without a doubt that biodiversity is dynamic and distributed equally on Earth, in general terms and over time.	3	14,29	3	13,64	100	100
Total	21	100	22	100		
Question 4 - This biodiversity is generally altered by environmental impacts. These, in turn, can be better defined as:						
A) Negative changes in the environment caused by human actions.	6	28,57	3	13,64	50	66,67
B) Major changes in the environment caused by human actions.	3	14,29	4	18,18	75	85,71
C) Negative and positive, large or small changes in the environment caused by human actions.	12	57,14	15	68,18	80	88,89
Total	21	100	22	100		
Question 5 - Another factor that can directly affect biodiversity in various places is global warming, which can currently be better defined as:						
A) A natural process that occurs on planet Earth and is a direct result of the rise in temperature caused by different factors.	0	0	2	9,09	0	0
B) A process caused directly by greenhouse gas emissions.	8	38,10	1	4,55	12,50	22,22
C) A natural process that occurs on planet Earth and is accelerated by greenhouse gas emissions.	13	61,90	19	86,36	68,42	81,25
Total	21	100	22	100		

Source: Direct research.

Among the alternatives, the concept that mentions biodiversity only as the diversity of living beings is simple and incomplete because it does not specify the various aspects linked to ecosystems. The second alternative, which mentions biodiversity as the association of various hierarchical components: ecosystem, community, species, populations and genes in a defined area, and the third alternative, which refers to biodiversity as the totality of genes, species and ecosystems in a region, are classed as the most complete definitions of biodiversity since they address the aspects mentioned in Article 2 of the Convention on Biological Diversity. In view of this, the percentage of students in classes A and B who chose these answers was relatively low. This may demonstrate a lack of in-depth knowledge of the term biodiversity. Despite this, it can be seen that the figures for class B show that there was an increase in the percentage of choices of the alternatives mentioned compared to class A.

Also in relation to the above question, the percentages of similarity were: 72.73% and 84.21% for the first alternative, 66.67% and 80% for the second, and 80% and 88.89% for the last, respectively for the Jaccard and Sorensen indices. When comparing the answers of classes A and B, it can be seen that the indices were close to each other, showing that there wasn't much difference in the choice of answers between the two classes.

For question 3, which asked which statement best characterised biodiversity in terms of the changes that could occur, we found that: 4.76% of the students in class A believe that biodiversity can be visualised as static, while no student in class B selected such an option. The alternative that mentions that biodiversity is immersed in a dynamic of constantly occurring changes was chosen by 80.95 per cent of the students in the first class (Class A) and 86.36 per cent in the other (Class B). The choice percentages for the last alternative were 14.29 per cent for Class A and 13.64 per cent for Class B. It mentions that biodiversity is dynamic and distributed equally on Earth, in general terms and over time (Table 2).

When considered in terms of change, biodiversity is actually immersed in a dynamic of changes that occur frequently, and which are not evenly distributed across the planet (BARBIERI, 1998). In view of this, it can be seen that most of the students in the two classes are aware of the dynamic nature of biodiversity transformation, i.e. its variability and mutability.

As for the similarity of the answers given, the indices showed high values with 89.47% and 100% in the Jaccard index and 94.44% and 100% in the Sorensen index for the second and third alternatives, respectively, of the question, while there was no similarity of answers in the first alternative, since this is the most incomplete and mistaken alternative in terms of biodiversity dynamics. This shows that there has been a slight change, since the graduating class, unlike the beginners, mostly opted for the most complete answer.

In question 4, the students were asked about the best definition that could be given to environmental impacts. In class A, 28.57 per cent of students believe that environmental impacts are negative changes to the environment caused by human actions; while 14.29 per cent say they are major changes to the environment caused by human actions; and 57.14 per cent say they are negative or positive, major or minor changes to the environment caused by human actions. In class B, the percentages collected, 13.64 per cent, 18.18 per cent and 68.18 per cent, correspond respectively to the alternatives mentioned above.

When we talk about environmental impact, we are clearly talking about changes to a greater or lesser degree, which may or may not contribute to maintaining the environment. So, given the above, most of the students chose the most complete answer.

Analysing the answers from the two classes, the similarity for the first alternative was 50% and 66.67%; the second 75% and 85.71%; and the third 80% and 88.89% for the Jaccard and Sorensen indices, respectively. These indices show that there was a regular difference in the choice of the first alternative (50% and 23%), and approximate differences of 25% and 14% in the second and 20% and 11% in the third alternative, with a higher percentage of class B opting for the latter.

Question 5 of the environmentalism sub-theme asks what the best definition of global warming is. The first alternative mentions global warming as a natural process occurring on planet Earth and the direct result of rising temperatures caused by different factors. Only a few students in class B chose this alternative, with a percentage equivalent to 9.09 per cent. The second option, which was chosen by 38.10 per cent of students in class A and 4.55 per cent in class B, mentions it as a process caused directly by greenhouse gas emissions. The last alternative, which states that global warming is a natural process occurring on planet Earth and that it is accelerated by greenhouse gas emissions, had the highest scores, with 61.90 per cent for class A and 86.36 per cent for class B.

In fact, there is currently no uniform definition of global warming. What does exist are lines of reasoning that differ from each other according to their own assumptions. According to the data collected on this question, it can be said that, due to the high percentages, the general thinking that revolves around a natural phenomenon that is directly influenced by human actions prevails in the majority.

With regard to the similarity indices, it can be seen that there is a huge difference in the choices between the two classes in question with regard to the first and second alternatives, which is reflected in the null or very low index. However, there is a higher percentage in the index for the third alternative, so there is a fair choice for this in both classes.

4.2 Environmentalism at UFPI

The second sub-theme deals with environmentalism at UFPI and has the same structure in terms of the number and types of questions.

As for the environmentalism sub-theme at UFPI, the only subjective question, which corresponded to the third question, asked students what measures UFPI should adopt in order to maintain and preserve the environment. There were a large number of answers, some of which expressed the same idea. The results obtained were then organised into two separate tables for each class: Table 3 and 4, for Class A and Class B.

The majority of students in class A, approximately 75%, believe that UFPI should carry out programmes and projects aimed at and encouraging environmental preservation, such as extension courses, as well as promoting campaigns on the institution's website and other campaigns. Other students, 10% of class A, mentioned that the institution should protect native species, maintain tree planting, improve the distribution of rubbish bins, establish awareness-raising projects, and carry out equipment maintenance. Among the students in this class, 5 per cent believe that UFPI should promote measures to create a group of students to promote the maintenance and preservation of the environment at UFPI. There were also those who didn't answer the question and didn't explain why. These corresponded to 10% of the class (Table 3).

Table 3: Measures that should be adopted by UFPI to maintain and preserve the environment according to undergraduates in Class A - Beginners.

Question 3 - In your opinion, what measures should UFPI adopt in order to maintain and preserve the environment?			
	Answers	N°	%
1	Carrying out programmes and projects aimed at and encouraging the preservation of the environment, such as extension courses, as well as promoting campaigns on this issue on the institution's website, in addition to other sites aimed at these campaigns.	15	75

		N°	%
2	Protecting native species, maintaining afforestation, improving the distribution of rubbish bins, establishing awareness-raising projects and maintaining equipment.	2	10
3	Promote the creation of a group of students to promote the maintenance and preservation of the environment at UFPI.	1	5
4	No answer	2	10
Total		20	100

Source: Direct research

So, according to the majority of students in this class, the main focus should be on creating projects aimed at preserving and conserving the environment, as well as publicising them.

Table 4 also deals with the same subject, i.e. the same question, but the answers were different, more diverse, and belonged to the students in class B.

Of the students in class A, 31.82% said that UFPI should include environmental education more effectively in the natural sciences and biology courses, as well as encouraging and supporting projects aimed at environmental education, such as debates and lectures on the sustainability of the countryside in an interdisciplinary way, covering all courses. The adoption of political preservation measures, encouraging rubbish collection, the maintenance and conservation of local fauna and flora, as well as lectures and mini-courses on the subject, also corresponded to 31.82% of Class B.

Other students, representing 9.09% of the class, mentioned that before carrying out any work, UFPI should survey the environmental impact and try to mitigate or compensate for it. It should also make all academic sectors aware of the importance and preservation of the environment, encouraging students to develop projects that can reduce environmental impacts.

Table 4: Measures that should be adopted by UFPI to maintain and preserve the environment, according to graduates of Class B.

Question 3 - In your opinion, what measures should UFPI adopt in order to maintain and preserve the environment?			
	Answers	**N°**	**%**
1	Introducing environmental education more effectively in the natural sciences and biology courses, as well as encouraging and supporting projects aimed at environmental education, such as debates and lectures, on the sustainability of the fields in an interdisciplinary way, covering all courses.	7	31,82
2	Preservation policies, encouraging rubbish collection, maintenance and conservation of local fauna and flora, as well as lectures and mini-courses on the subject.	7	31,82
3	Before any work is carried out, UFPI should survey the environmental impact and try to mitigate or compensate for it. Make all academic sectors aware of the importance and preservation of the environment. Encourage students to develop projects that can reduce environmental impacts.	2	9,09

4	Provide some kind of preparatory course for outsourced workers who deal with rubbish collection and general services, so that they are better prepared to handle selective and non-selective rubbish collection, in order to prevent rubbish dumps from increasing on the estate.	1	4,55
5	UFPI could adopt educational campaigns on the subject in partnership with biology students. In addition, a subject dealing with the issue should be included in the curriculum of all courses.	2	9,09
6	Planning to verticalise the campus, preserving the areas that have not yet undergone human intervention is an alternative that should be considered by managers.	1	4,55
7	It should encourage the development of projects, create partnerships with various schools in Teresina and the interior of Piauí in order to bring ideas for improving the use of natural space.	1	4,55
8	I don't have any measures in mind.	1	4,55
Total		22	100

Source: Direct research

The option of the institution providing a preparatory course for outsourced workers who deal with rubbish collection and general services, with a view to better preparing them to handle selective and non-selective collection, was mentioned by 4.55% of the students. This percentage was the same for those students who said that the UFPI should plan to verticalise the UFPI Campus and preserve the areas that have not yet undergone human intervention; and also for those who said that there should be incentives to develop projects, create partnerships with various schools in Teresina and the interior of Piauí so that there is an exchange of ideas aimed at improving the use of natural space.

Next, 9.09% of Class B said that UFPI could adopt educational campaigns on the subject in partnership with biology students, as well as inserting a subject dealing with environmental preservation and maintenance into the curriculum of all courses. It is worth noting that 4.55% of these students did not mention any measures.

Among the objective questions, question 1 asks, in the students' view, what level of concern UFPI, as an institution, has for maintaining and preserving the environment; question 2 asks whether UFPI should adopt projects aimed at preserving the environment and also disseminating environmental knowledge; question 4 asks what is the best alternative that undergraduates as a group should seek in order to disseminate environmental knowledge; and question 5 asks what are the best benefits generated by the implementation by UFPI of a set of measures aimed at maintaining and preserving the environment.

Of the three alternatives in question 1, the first was 4.76% and 0%, the second was 80.95% and 50%, and the third was 14.29% and 50%, respectively.

Analysing the indices for the two classes, it can be seen that there was no similarity in the choice of the first alternative, since the percentage was very low for

Class A and zero for Class B. As for the choice of the second item, there was a percentage of 64.71% and 78.57%, respectively for Jaccard and Sorensen, which shows that the students believe that the level of concern at UFPI is regular, especially for those in the beginner class. In the third alternative, there were low similarity values, 27.27% and 42.86%, respectively for the Jaccard and Sorensen indexes, which indicates that there was a notable difference in the choice of this alternative between the classes. This is due to the fact that Class B is much more likely to believe that UFPI's concern with maintaining and preserving the environment is bad (Table 5).

Table 5. Environmentalism at UFPI according to undergraduates at the beginning and end of the Biological Sciences degree course at UFPI.

Question 1 - In your opinion, how concerned is UFPI as an institution about maintaining and preserving the environment?	Class A		Class B		similarity index	
	N°	%	N°	%	Jaccard	Sorensen
A) Good, because UFPI has a set of measures aimed at this approach that have produced good results.	1	4,76	0	0	0	0
B) Regular, because the measures taken by the UFPI are not exactly sufficient, but they help in this approach.	17	80,95	11	50	64,71	78,57
C) Bad, because none of the measures are effective or there are no measures in place at UFPI for this approach.	3	14,29	11	50	27,27	42,86
Total	21	100	22	100		
Question 2 - Should UFPI adopt projects aimed at preserving the environment and disseminating environmental knowledge?						
A) Yes.	21	100	22	100	95,45	97,67
B) No.	0	0	0	0	-	-
C) Maybe.	0	0	0	0	-	-
Total	21	100	22	100		
Question 4 - What about undergraduates as a group, what is the best alternative they should look for in order to disseminate environmental knowledge?						
A) Organising events such as environmental week and/or displaying posters with messages about this approach.	4	19,05	3	13,64	75	85,71
B) Dissemination of science fairs in the block.	0	0	0	0	-	-
C) Symposiums, mini-curricula, lectures, regular spaces for developing work, developing projects in this approach, etc.	17	80,95	19	86,36	89,47	94,44
Total	21	100	22	100		
Question 5 - In your opinion, what are the best benefits generated by UFPI's implementation of a set of measures aimed at maintaining and preserving the environment?						
A) Raising awareness among students, teachers, technicians and others.	8	38,10	6	28,57	75	85,71
B) The consequent preservation of the environment and the sustainable use of spaces at UFPI, in terms of buildings.	8	38,10	5	23,81	62,50	76,92
C) The maintenance of green areas and the reduction of environmental damage at UFPI.	5	23,81	10	47,62	50	66,67
Total	21	100	21	100		

Source: Direct research

Question 2 asked whether UFPI should adopt projects aimed at preserving the environment and disseminating environmental knowledge. The options were "yes", "no" and "maybe", however, the choices were totally similar and unanimous, with the students opting for the first alternative, "yes". In other words, both Class A and Class B students agree that the measure described by the institution really should be taken (Table 5).

Question 4 asks what the best alternative should be for undergraduates as a group to disseminate environmental knowledge. Of the alternatives, the first mentions organising events such as environmental week and/or displaying posters with messages about this approach. This was chosen by 19.05% of the beginners and 13.64% of the graduates. The second option mentions the dissemination of science fairs in the block, but the percentages were zero, i.e. none of the students in the two classes chose this. The vast majority of students opted to seek out symposiums, mini-courses, lectures, regular spaces for developing work, developing projects in this approach, among others.

80.95 per cent and 86.36 per cent for Class A and Class B, respectively (Table 5).

According to the Jaccard and Sorensen indices, there were high levels of similarity in the choice of answers: 75 per cent and 85.71 per cent for the first alternative, 89.47 per cent and 94.44 per cent for the third alternative. In other words, the two classes did not differ greatly in their choices, and therefore expressed equal opinions. As no student opted for the second alternative, no real values were obtained for the similarity index (Table 5).

In relation to question 5, which is part of the environmentalism sub-theme at UFPI, the students were asked about the best benefits generated by UFPI's implementation of a set of measures aimed at maintaining and preserving the environment. In view of this, 38.10 per cent of the students in Class A and 28.57 per cent of the students in the concluding class said that raising the awareness of undergraduates, teachers, technicians and other people was the best benefit, while 38.10 per cent of the beginning class and 23.81 per cent of Class B believed that it was the preservation of the environment and the sustainable use of spaces at UFPI, in terms of buildings. Maintaining green areas and reducing environmental damage at UFPI was chosen by 23.81 per cent of students in the beginners' class and 47.62 per cent in the graduates' class (Table 5).

In terms of similarity, there were 75% and 85.71% for the first alternative; 62.50% and 76.92% for the second; and 50% and 66.67% for the third alternative, the indices coming from Jaccard and Sorensen respectively. It can therefore be seen that there was not a great difference in the choices of the first two options by the two classes, however, the third alternative showed a greater difference, as a higher percentage of Class B students chose it (Table 5).

4.3 Selective collection

The third sub-theme focused on the subject of selective collection, again along the lines presented: four objective questions and one subjective question. The subjective question asked the students what concept they would attribute to selective collection, and the results are shown in Table 6.

Table 6. Concepts attributed to selective collection according to undergraduates from the initial and final periods of the Biological Sciences degree course at UFPI.

Question 1-0 What is selective waste collection?		Class A		Class B		similarity index	
Answers		N°	%	N°	%	Jaccard	Sorensen
1	It is the act of collecting and separating rubbish when it is discarded with a view to reusing or recycling it, thus contributing to the preservation of the environment and its sustainability.	10	47,62	14	63,64	71,43	83,33
2	It is the separation of rubbish according to its chemical composition.	9	42,86	8	36,36	88,89	94,12
3	Household rubbish collection.	1	4,76	0	0	0	0
4	He didn't answer.	1	4,76	0	0	0	0
Total		21	100,00	22	100,00		

Source: Direct research.

It is worth noting that various concepts were attributed to selective waste collection, but it was decided to organise them according to the idea they expressed, resulting in three distinct concepts for both the beginners' and the graduates' classes.

There were four distinct categories among the answers. The most complete answer mentioned by the students was that selective collection was the act of collecting and separating rubbish when it was discarded, with a view to reusing or recycling it, thus leading to the preservation and sustainability of the environment. In Class A, the percentage of students who mentioned this option was 47.62 per cent, while in Class B this figure was 63.64 per cent. Given that the students in Class B had a higher percentage for this option, it could be proposed that they had a better idea of the concept attributed to selective collection. The similarity between the answers for this alternative was high, with 71.43 per cent for Jaccard and 83.33 per cent for Sorensen, indicating that there was a similar number of students in the two classes who opted for this statement (Table 6).

Another answer that also showed high rates was the one that mentioned that selective collection was about separating rubbish according to its chemical composition. Compared to the other answer, this one is less complete, as there is no mention of reusing rubbish. For this alternative, 42.86% of Class A and 36.36% of Class B affirmed it. The fact that Class B had a lower percentage for this statement is in line with what was mentioned when analysing the first alternative. The similarity indices were 88.89% and 94.12% for Jaccard and Sorensen, respectively. These values were high for the same reason as mentioned in the previous option (Table 6).

The third alternative mentioned that selective collection was nothing more than the collection of household rubbish. This concept is very incomplete, with, for example, no specificity in the approach to rubbish collection. In this case, 4.76 per cent of class A chose to affirm it, while there were no students in class B who affirmed this concept for selective collection. As a result, there was no similarity (Table 6).

On this question, 4.76 per cent of class A did not answer the question, while all the students in class B did (Table 6).

Among the objective questions, question 2 asks students what the correct colours are for each specific type of waste; question 3 asks which method or methods used to reuse or process waste would be the most viable for the environment; question 4 asks what could best be attributed the great importance of selective collection; and question 5 asks what factors could most hinder the success of selective collection (Table 7).

Table 7. Selective collection according to undergraduates in the initial and final periods of the Biological Sciences degree course at UFPL.

Question 2 - The prior separation of the various types of rubbish and waste begins with the waste containers, which have different colours for each type of rubbish. With regard to the colours of waste containers, the following can be said:	Class A		Class B		similarity index	
	N°	%	N°	%	Jaccard	Sorensen
A) The colour green is for plant materials, blue for plastics and yellow for paper / cardboard.	5	23,81	0	0	0	0
B) Brown is for organic waste, green for glass and white for hospital/healthcare waste.	6	28,57	17	80,95	35,29	52,17

C) The colours purple, orange and grey are not used for waste containers.	10	47,62	4	19,05	40	57,14
Total	21	100	21	100		
Question 3 - Among the methods used to reuse or process rubbish, which of the following is the most environmentally viable?						
A) Landfill, where rubbish is deposited under layers of earth and the methane gas is collected or incinerated.	3	14,29	1	4,55	33,33	50
B) The incineration of rubbish, which drastically reduces the space it takes up in the environment.	0	0	0	0	-	-
C)Recycling and composting to reuse rubbish.	18	85,71	21	95,45	85,71	92,31
Total	21	100	22	100		
Question 4 - The great importance of selective waste collection can best be attributed to:						
A) Reducing costs for the public purse, which consequently doesn't have to spend so much looking for new spaces to build landfill sites or rubbish dumps.	1	4,76	0	0	0	0
B) Benefit to the environment, as separating rubbish directly enables the recycling process, as well as reducing the volume of rubbish for final disposal.	18	85,71	22	100	81,82	90
C) Facilitating the work of recyclers and waste collectors.	2	9,52	0	0	0	0
Total	21	100	22	100		
Question 5 - The factors that can hinder the success of selective collection are best represented by:						
A) The lack of structures or places to process rubbish separately (recycling and composting).	5	23,81	6	27,27	83,33	90,91
B) Only the government's own lack of interest in drawing up plans for the success of selective collection.	2	9,52	1	4,55	50	66,67
C) Lack of public awareness and government incentives.	14	66,67	15	68,18	93,33	96,55
Total	21	100	22	100		

Source: Direct research.

Question 2 of this theme, when asking students about the correct colours for each specific type of waste, proposes three alternatives, the first of which is green for the disposal of plant materials, blue for plastics and yellow for paper, cardboard; the second, that brown is used for organic waste, green for glass and white for hospital/healthcare waste; and the third, that purple, orange and grey are not used for waste containers. For the first alternative, there was a percentage of 23.81 per cent for Class A, while no students in Class B chose this alternative. For the second alternative, the values were 28.57% and 80.95%, respectively for the beginners and final year classes. For the last option, the figures were 47.62% for Class A and 19.05% for Class B (Table 7).

The similarity index between the options in this question showed low values, with the second and third options respectively showing 35.29% and 52.17%, and 40% and 57.14%, in line with the two indices that have been discussed. In the first option, there was no similarity at all, as no student in the graduating class chose this alternative. This disparity is mainly due to the fact that the students in Class A showed greater knowledge of the selective collection process in terms of the colours used to identify specific waste containers (Table 7).

Question 3 asked which method or methods used to reuse or process rubbish would be the most environmentally viable. In Class A, 14.29% answered that landfill, where rubbish is deposited in alternating layers with layers of earth and methane gas is collected or incinerated, was the best method, while in Class B only 4.55% chose the latter. The second alternative mentioned the incineration of rubbish, which drastically reduces the space it takes up in the environment, as the most viable way, but this was not selected by any student in either class. In the case of the third alternative, which stated that recycling and composting are the best ways to reuse rubbish, the majority of students chose it, with 86.71% of students in the beginners' class choosing it and 95.45% of students in the other class choosing it (Table 7).

As for the similarity of choices related to the question being addressed, the Jaccard index showed only 33.33% and 85.71% between the two classes for the first and third alternatives, respectively. For the Sorensen index, the respective percentages were 50 per cent and 92.31 per cent. It can also be seen that the fact that no student opted for the second alternative emphasises the impossibility of having real values for the similarity index.

Analysing all the alternatives in the third question leads to the conclusion that the best way to reuse and process waste without causing major problems for the environment is through recycling and composting. These methods not only reduce the space taken up by waste, but also make it useful again, i.e. ready for reuse. Given that the majority of students chose the correct alternative, this shows that they have a basic knowledge of how to dispose of rubbish. The other two methods are not the best alternatives because landfill, even though there is a whole system in place to prevent soil contamination, requires a large area for rubbish disposal, which in turn will come into contact with the overlying soil, leaving that space unviable for other activities. And incineration, although it reduces the volume of rubbish and kills micro-organisms, produces a large quantity of toxic gases for the environment and requires very expensive treatment methods.

Question 4 asked what could best be attributed to the great importance of selective collection. Of the options, the first was the reduction in spending by the public purse, which would then not have to spend so much looking for new spaces to build landfill sites or rubbish dumps. Only 4.76 per cent of the students in the beginners' class chose this alternative, while none of the students in the graduates' class did. The second alternative mentioned the benefit to the environment, since separating rubbish directly enables the recycling process, as well as reducing the volume of rubbish for final disposal. The percentages for this choice were higher, as 85.71 per cent of Class A chose it, while all the students in Class B chose it. The alternative that portrayed facilitating the work of recyclers, as well as people who work in rubbish collection, as the reason why selective collection is so important, was chosen by 9.52 per cent of Class A students, while none of Class B chose this option.

The similarity indices in the choice of answers by the two classes were zero for the first and third alternatives, i.e. there were no common answers. This is due to the fact that all the students in Class B chose only the second alternative. This, in turn, showed rates of 81.82 per cent and 90 per cent, respectively for Jaccard and Sorensen. The students did well because most of them chose the second alternative and it was the right one. However, in terms of what was said, it can be seen that the graduating class was much more aware of the importance of selective collection than the beginners (Table 7).

Finally, question 5 of the aforementioned sub-theme asks which factors could most hinder the success of selective collection. Of the options, the first had similar rates between the two classes, with 23.81 per cent of Class A students choosing it and 27.27 per cent of Class B students choosing it. Also related to this alternative, the similarity indices were 83.33% and 90.91%, respectively for Jaccard and Sorensen, indicating that some students in both classes chose the same alternative, since this cited the lack of structure or places to process waste separately as an obstacle to selective collection. The factors cited by this alternative do indeed interfere with the selective collection process, but when analysing it with the other options, it can be seen that it does not include the factors that would most influence what has been said.

The second option stated that the reason for the failure of selective collection was simply the government's own lack of interest in drawing up plans for its success. Even though this alternative is clearly an incomplete and mistaken choice due to the limiting factor "only", 9.52 per cent of the beginner class and 4.55 per cent of the graduate class chose it. The similarity indices for this alternative were 50 per cent for Jaccard and 66.67 per cent for Sorensen, which is why the graduating students chose it less. The third and most complete alternative was chosen by 66.67 per cent of the beginners and 68.18 per cent of the graduates. It mentioned that the factors that most influenced the lack of success of selective collection were people's lack of awareness and the government's lack of incentives. As the percentages were similar, within a context where the number of survey participants for the two classes was almost the same, the similarity indices for this option were high, at 93.33% and 96.55% for Jaccard and Sorensen, respectively. This indicates that the majority of students are aware of the mechanisms needed for selective collection to work.

4.4 Graduation

Finally, the last theme, which was divided into questions, addressed some basic aspects related to environmental knowledge and the graduation of the students in the two classes. Question 4 asked which environmental knowledge or concepts had been assimilated before entering the Biological Sciences undergraduate course for students in class A, and during the Biological Sciences undergraduate course for students in class B. The question asked which environmental knowledge or concepts had been assimilated before entering the Biological Sciences undergraduate course. As a greater number of different and similar concepts were mentioned by the students in relation to question 4, we decided to calculate their rate of occurrence, resulting in two graphs: Graph 1 for Class A and Graph 2 for Class B.

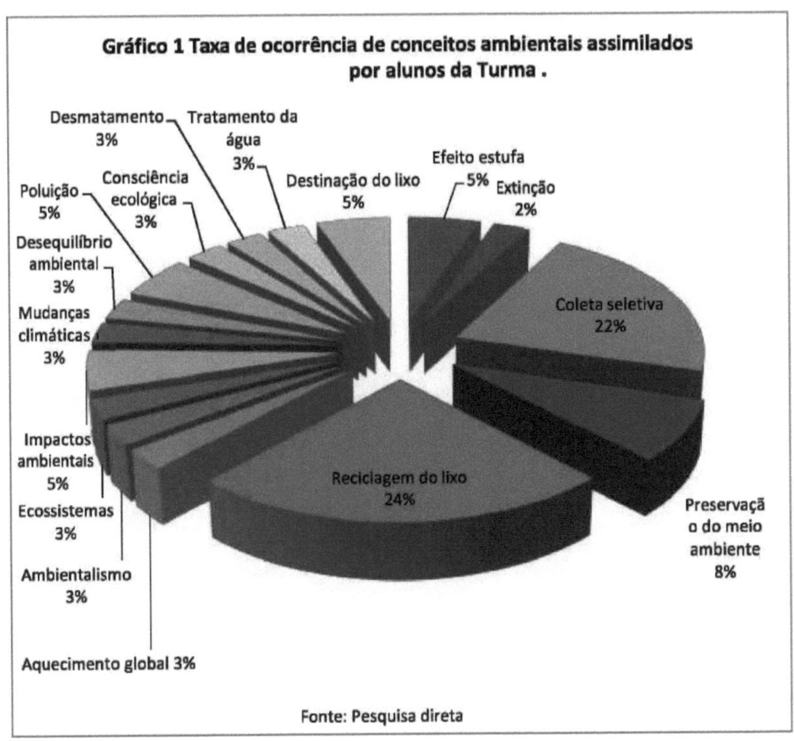

For Class A, the 16 assimilated concepts most cited by the students were: rubbish recycling, with a percentage of approximately 22%; selective collection, with 20%; environmental preservation, with 7%; as well as environmental impact, greenhouse effect, rubbish disposal and pollution, with rates equal to 5%. In Class B, as shown in Graph 2: Rate of occurrence of concepts assimilated by Class B students, there were a greater number of concepts presented with very different percentages.

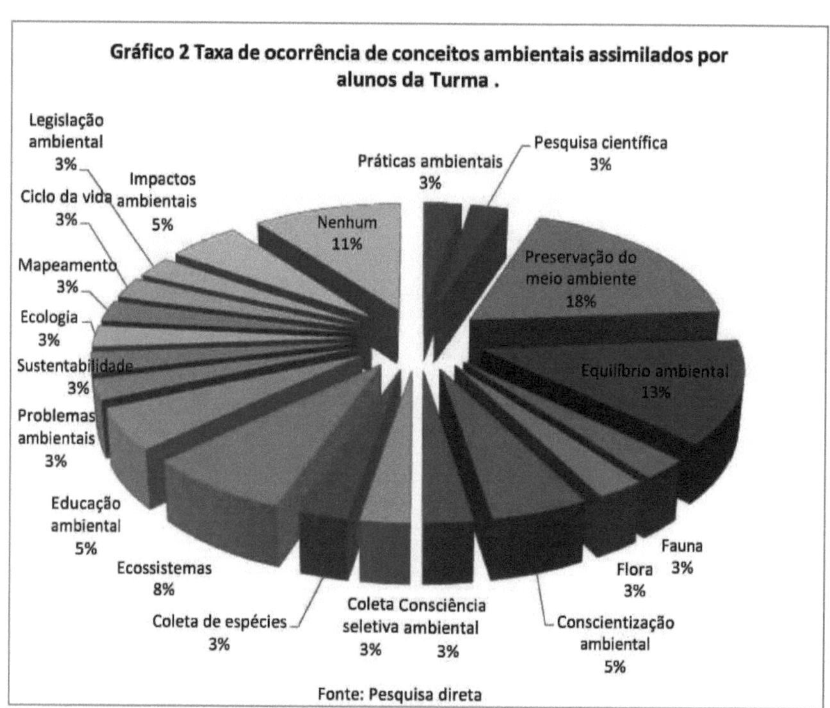

In this class, of the 19 assimilated concepts mentioned, the ones with the highest occurrence rates were environmental preservation, with 18 per cent; environmental balance with 13 per cent; and ecosystems with 8 per cent. Concepts such as environmental awareness, environmental education and environmental impacts had a 5% occurrence rate. Other concepts had an occurrence rate of 3%. There were also those students who said they had not assimilated any specific concept, which accounted for 11% of the occurrence.

As for the similarity between the concepts presented by the two classes, we have values of 13 per cent for the Jaccard index and 23 per cent for the Sorensen index. This relatively low similarity is due to the fact that there were few common terms mentioned by the two classes.

Question 1 of the aforementioned sub-theme was different for the two classes. In the beginners' class, this question asked how the students saw their entry into the Biological Sciences degree course in relation to environmental knowledge.

Table 8. Graduation, specific approach according to undergraduates from the initial and __________ final periods of _____ the Biological Sciences degree course at UFPI.

Classes	Question 1 - How do you see your entry to the degree course in biological sciences in relation to environmental knowledge?	N°	%
A	As an opportunity to learn and improve my environmental knowledge.	17	80,95
	I don't think my environmental knowledge can be expanded.	0	0
	I'm not sure what benefits this course can bring me in terms of environmental knowledge.	4	19,05
	Total	21	100
	Question 1 - What did the degree course in biological sciences bring you in terms of environmental knowledge?		
B	It gave me the opportunity to learn and improve on what I already knew.	10	45,45
	It brought me completely new information, knowledge and content.	6	-2.1,21
	It didn't give me any new knowledge.	6	27,27
	Total	22	100

Source: Direct research.

80.95% of the students in class A said it was an opportunity to learn and improve their environmental knowledge, while only 19.05% of them were not sure what benefits the Biological Sciences degree course could bring in terms of environmental knowledge. For the second alternative, which mentioned that the students did not believe that their environmental knowledge could be expanded, the percentage of choices was nil. The choice of these alternatives may indicate that the students in this class understand the changing nature of knowledge, although some do not imagine such a possibility (Table 8).

As for class B, question 1 asked what the degree course in biological sciences had brought them in terms of environmental knowledge. The figures were better distributed among the alternatives, with 45.45 per cent of the students saying that the course had provided learning and improvement on what they already knew; 27.27 per cent of the students saying that the course had provided completely new information, knowledge and content; and 27.27 per cent of them saying that the course had not provided any new knowledge. With these percentages, it can be seen that, to a certain extent, for most of the students in class B, the course itself brought new learning.

Unlike question 1 and question 4, the other questions involving the graduation sub-theme were the same for both classes.

Question 2 asked the students whether the subjects they were offered that dealt with issues related to environmental knowledge were sufficient, insufficient or if they didn't have an opinion. In this regard, 42.96% of Class A and 45.45% of Class

B said that the subjects on offer were sufficient, but that they needed to improve their approaches to environmental issues. On the other hand, 33.33% of Class A students and 45.45% of Class B students said that the subjects were insufficient because they didn't address the environmental aspect in any understandable or direct way. Of the total number of students in Class A and Class B, only 23.91% and 9.09%, respectively, said they didn't have an elaborate concept (Table 9).

With regard to the similarity of the answers, the Jaccard and Sorensen scores for the first alternative were 90 per cent and 94.74 per cent respectively, indicating that the two classes had similar choices when it came to the subjects being sufficient. In the second alternative, the rates were 70% and 82.35%. In the third alternative, the similarity indices were 40 per cent and 57.14 per cent, indicating that there was a big difference in the choice for this alternative. This difference, in a comparison between the two classes, can be understood by the fact that most of the students in class B have an elaborate conception.

Question 3 asked students who should demand that the various aspects of environmental knowledge be addressed. In class A, 15 per cent of students said that it was the teachers and students themselves who would demand what was mentioned, while in class B, 13.64 per cent chose the same alternative. Those who said that the course coordinator should make this demand were 10% of class A and 4.55% of class B. The percentages of 75 per cent for class A and 81.82 per cent for class B corresponded to those students who said that everyone (students, teachers and coordination) should demand what was said (Table 9).

The similarity in the answers to the first alternative in this question was 100 per cent, i.e. the same number of students chose it. For the second, the Jaccard and Sorensen indices were 50% and 66.67%. This similarity was lower because the students in class B were much more likely to believe that the requirement should be made by everyone. The similarity in responses was high for the third alternative, with indices equal to 83.33% for Jaccard and 90.91% for Sorensen, because, in general, the majority of students believe in this option.

Table 9. Graduation, general approach according to undergraduates from the initial and final periods of the

Biological Sciences degree course at LIFP.

Question 2 - In your opinion, the subjects on offer that deal with themes linked to environmental knowledge are:	Class A		Class B		similarity index	
	N°	%	N°	%	Jaccard	Sorensen
A) Sufficient, but they need to improve their approach to this environmental aspect.	9	42,86	10	45,45	90	94,74
B) Insufficient, as they do not address this environmental aspect in any understandable or direct way.	7	33,33	10	45,45	70	82,35
C) I don't have an elaborate concept.	5	23,81	2	9,09	40	57,14
Total	21	100	22	100		
Question 3 - In your opinion, who should be required to address the various aspects of environmental knowledge?						
A) By the teachers and students themselves.	3	15	3	13,64	100	100
B) By the course coordinator.	2	10	1	4,55	50	66,67
C) For everyone.	15	75	18	81,82	83,33	90,91
Total	20	100	22	100		
Question 5 - In your opinion, your level of performance in the search for environmental knowledge is:						
A) Good, because I try to keep up to date with this aspect.	10	47,62	11	50	90,91	95,24
B) Fair, because it's a theme that catches my attention, but doesn't hold it that much.	11	52,38	10	45,45	90,91	95,24
C) Bad, because I'm not usually interested in these approaches.	0	0	1	4,55	0	0
Total	21	100	22	100		

Source: Direct research

Question 5 analysed the students on their own conception of their level of performance in the search for environmental knowledge. Of the beginner class, 47.62 per cent said that their level of performance was good, as they sought to find out or update themselves on this aspect, while half of the graduating class said the same. The majority (52.38%) of the students in the beginners' class chose to say that their level of performance was fair, because this topic caught their attention, but they didn't look into it as much. This option was chosen by 45.45% of the graduating class. As for the statement that their level of performance in the search for environmental knowledge was poor, because they had no interest in such an approach, only a minority of 4.55% of class B said this, while none of the students in class A chose this option (Table 9).

The similarity indices were high and equal for the first and second alternative, with 90.91 per cent and 95.24 per cent respectively for Jaccard and Sorensen. This indicates that there were not many changes in the students' opinions between the two classes. In the third alternative, there was no similarity between the answers from the two classes, precisely because only one student, from class B, chose this option.

CHAPTER 5

FINAL CONSIDERATIONS

The need to develop environmental awareness makes it even more essential to disseminate knowledge that is mainly related to the practice of preserving and maintaining the environment. The main means of achieving this is through environmental education, which is applied a priori in teacher training.

In view of the objectives proposed by this work, it was possible to observe the extent to which undergraduate students in biological sciences master the various aspects of environmental knowledge, their interest in this subject, and their own satisfaction with the course curriculum, also in this approach. By analysing the data between the two classes (the initial period class and the final period class), it was also possible to relate the possible influence of the course on the formation of these students' environmental knowledge.

It can therefore be understood that the students generally showed good mastery, since the answers they gave or chose that were actually more complete or even more correct in relation to what they were saying achieved high percentages of occurrence most of the time. It is also worth pointing out that there was a difference in the level of understanding of the environmental theme between the two classes, with the class that was in its final periods showing greater mastery of environmental knowledge than the other. One hypothesis that could be attributed to this difference would be the influence of environmental approaches in the subjects throughout the course, as well as the students' own maturity.

With regard to interest in environmental issues, it can be seen that the majority of students have a regular interest. This can be seen when they assess the level of their quest to develop environmental knowledge, and also when they mention the measures that should be adopted to preserve and maintain the environment at UFPI. From this, too, it can be seen that the students are not satisfied with the measures taken by the institution, not least because they cited several others that it should introduce. Another crucial point for the research is that among the students interviewed, most of them believe that the approach to environmental knowledge in the various disciplines of the course should be improved, to the point of making it easier to understand. It is hoped that all the data collected in this research can serve as an influence for other studies, as well as for taking measures aimed at preserving and maintaining the environment; and aimed at improving the approach to environmental issues, especially for courses aimed at training teachers.

CHAPTER 6

REFERENCES

BARBIERI, E. **Biodiversity:** green capitalism or social ecology? São Paulo: Editora Cidade Nova, 1998. 89p.

Brazil, MMA. **The Convention on Biological Diversity -** CBD, Copy of Legislative Decree No. 2 of 5 June 1992. MMA. Brasilia, 2002. 30 p.

CARVALHO, I. C. de M. **Educação ambiental:** a formação do sujeito ecológico. 3ª ed. São Paulo: Cortez, 2008.

CUNHA, S. B. da (Org.); GUERRA, Antonio José T. (Org). **The environmental question:** different approaches. 3ª ed. Rio de Janeiro: Bertrand Brasil, 2007. 248 p.

FLORIANI, D. **Knowledge, environment & globalisation.** 1ª ed (year 2004). 4ª tir. Curitiba: Juruá, 2008. 174 p.

GUIMARÃES, M. **A dimensão ambiental na educação.** 10ª ed (2010). Capinas: Papirus, 1995. 96 p.

LOUREIRO, C. F. B.; LAYRARGUES, P. P.; CASTRO, R. Souza de (eds). **Society and the environment:** environmental education in debate. 4ª ed. São Paulo: Cortez, 2006.

MATTEUCCI, S. D; COLMA, A. **Metodologia para ei estudo de la vegetacion.** Washington, OAS/PRDECT, 168 p. 1982.

MUELLER-DOMBOIS, D; ELLENBERG, H. **Aims and methods vegetation ecology.** New York: John Wiley and Sons, 1974. 547 p.

RAMOS, O. C.; COSTA, M. D. B. **Ecologia e meio ambiente.** Goiânia: Brasília jurídica LTDA, 1992. 1546 p.

SANTOS, A. M. O. dos. **A guided tour as a teaching technique in environmental education.** 2013. 47 f. Monograph (Undergraduate degree in biological sciences) - Nature Sciences Centre, Federal University of Piauí, Teresina.

SILVESTRE, R. **Comparison of floristics, structure and spatial pattern in three fragments of mixed ombrophilous forest in the state of Paraná.** 89 f. Dissertation (Master's in Forestry Sciences) - Agrarian Sciences Sector, Federal University of Paraná, Curitiba, 2009.

CHAPTER 7

ANNEXES

INFORMED CONSENT FORM

Title of the study: Diagnosis of the environmental knowledge of graduate biologists in Biological Sciences at UFPI.
Researcher responsible: Lucas Soares da Silva
I nstitution/ Departm ent: UFPI/ Departm ent of Biology
Contact e-mail: lucas_soares120@hotmail.com
Place of data collection: UFPI Dear Graduating Student:

You are being invited to answer the questions in this questionnaire on a completely **voluntary basis.** Before you agree to take part in this research and answer this questionnaire, it is very important that you understand the information and instructions contained in this document.

Aim of the study: To diagnose the state of graduate biologists' knowledge of basic and more complex concepts in the environmental field, at the "Ministro Petrônio Portela" campus of UFPI in Teresina, covering two undergraduate classes: one in the first term and the other in the last term of the Biological Sciences Undergraduate Course.

Procedures. Your participation in this survey will only consist of filling in this questionnaire, answering the questions that address adverse issues related to environmental knowledge.

Benefits. This research will provide greater knowledge on the subject, and the results collected may subsequently lead to improvements in the biological sciences degree course itself.

Risks. Completing this questionnaire will not pose any physical or psychological risk to you.

Confidentiality. The privacy of the information you provide will be guaranteed by the researchers responsible. Research subjects will not be identified at any time, even when the results of this research are publicised in any form.

Aware of and in agreement with the above, I, under the registration number, agree to take part in this research.

____
Researcher responsible

CLASS A QUESTIONNAIRE

This questionnaire is for research purposes only. All the data collected will be used in a final year coursework. The questionnaire has a total of 20 (twenty) questions grouped into various themes and distributed between subjective and objective questions. In the objective questions, **only one option** should be ticked out of the three proposed.

ENVIRONMENTALISM

1. **How would you describe the environment?**

____

____

____

____

2. According to your knowledge, the term biodiversity can best be described as:
( ) Diversity of living beings.
( ) Association of various hierarchical components: ecosystem, community, species, populations and genes in a defined area.
( )The totality of genes, species and ecosystems in a region.

3. What can we say about the changes that could occur to this biodiversity?
( ) Biodiversity can be seen as static.
( ) Biodiversity is immersed in a dynamic of changes that occur all the time.
( ) It can be said without a doubt that biodiversity is dynamic and distributed equally on Earth, in general terms and over time.

4. This biodiversity is usually altered by environmental impacts. These, in turn, can be better defined as:
( ) Negative changes in the environment caused by human actions.
( ) Major changes in the environment caused by human actions.
( ) Negative and positive, large or small changes in the environment caused by human actions.

5. Another factor that can directly affect biodiversity in various places is global warming, which can currently be better defined as:
( ) A natural process that occurs on planet Earth and is a direct result of the rise in temperature caused by different factors.
( ) Process caused directly by greenhouse gas emissions.
( ) A natural process that occurs on planet Earth and is accelerated by greenhouse gas emissions.

ENVIRONMENTALISM AT UFPI

1 . In your opinion, how concerned is UFPI as an institution about maintaining and preserving the environment?
( ) Good, because UFPI has a set of measures aimed at this approach that have had good results.
( ) Fair, because the measures taken by the UFPI are not exactly sufficient, but they do help with this approach.
(   ) Bad, because none of the measures are effective or there are no measures in place. UFPI for this approach.

2 Should UFPI adopt projects aimed at preserving the environment and disseminating environmental knowledge?
( ) Yes   ( ) No   ( ) Maybe

1 In your opinion, what measures should UFPI adopt in order to maintain and preserve the environment?

____

____

4 What about undergraduates as a group, what is the best alternative they should look for in order to disseminate environmental knowledge?

(   ) Organising events such as environmental week and/or display posters with messages about this approach.

( ) Dissemination of science fairs in the block.

( ) Symposiums, mini-courses, lectures, regular spaces for developing work, developing projects in this approach, etc.

& In your opinion, what are the best benefits generated by UFPI's implementation of a set of measures aimed at maintaining and preserving the environment?

( ) Raising awareness among undergraduates, teachers, technicians and others.

( ) The consequent preservation of the environment and the sustainable use of spaces at UFPI, in terms of buildings.

( ) Maintaining green areas and reducing environmental damage at UFPI.

SELECTIVE COLLECTION

1. What is selective collection?

2. Prior separation of the various types of rubbish and waste begins with the waste containers, which are coloured differently for each type of waste.

rubbish. With regard to the colours of waste containers, we can say:

( ) The colour green is for plant materials, blue for plastics and yellow for paper / cardboard.

(   ) The colour brown is for organic waste, green is for glass and white paper for hospital/healthcare waste.

(   ) Purple, orange and grey colours are not used in containers waste.

3. Among the methods used to reuse or process rubbish, which of the following is the most environmentally friendly?

(   ) Landfill, where rubbish is deposited under layers of earth with collection or incineration of methane gas.

( ) The incineration of rubbish, which drastically reduces the space it takes up in the environment.

( ) Recycling and composting to reuse rubbish.

4. The great importance of selective collection can be better attributed to: ( ) Reducing costs for public coffers, which consequently don't have to spend so much looking for new spaces to build landfills or rubbish dumps.

( ) Benefit to the environment, as separating rubbish directly enables the recycling process, as well as reducing the volume of rubbish for final disposal.

( ) Facilitating the work of recyclers, as well as people who collect rubbish.

5. The factors that can hinder the success of selective collection are best represented by:
( ) The lack of structures or places to process rubbish separately (recycling and composting).
(   ) Only a lack of interest on the part of the government itself in drawing up plans for the success of selective collection.
( ) Lack of public awareness and government incentives.

GRADUATION

6. How do you see your entry into the degree course in biological sciences in relation to environmental knowledge?
( ) As an opportunity to learn and improve my environmental knowledge.
( ) I don't think my environmental knowledge can be expanded.
( ) I'm not sure what benefits this course can bring me in terms of environmental knowledge.
7. In its conception, the subjects on offer, which deal with themes linked to environmental knowledge, are:
( ) Sufficient, but they need to improve their approaches to this environmental aspect.
( ) Insufficient, as they do not address this environmental aspect in any way

understandable or direct.
( ) I don't have an elaborate concept.
3. In your opinion, who should demand that the various aspects of environmental knowledge be addressed?
( ) By the teachers and students themselves.
( ) By the course coordinator.
( ) For everyone.
4. What environmental knowledge did you acquire before you started your degree course in biological sciences?

____
____
____
____

5. In your opinion, your level of performance in the search for environmental knowledge is:
( ) Good, because I try to find out about it or keep up to date with it.
( ) Fair, because it's a topic that catches my attention, but doesn't hold it that much.
( ) Bad, because I'm not usually interested in these approaches.

CLASS B QUESTIONNAIRE

This questionnaire is for research purposes only. All the data collected will be used in a final year coursework. The questionnaire has a total of 20 (twenty) questions grouped into various themes and distributed between subjective and objective questions. In the objective questions, **only one option** should be ticked out of the three proposed.

ENVIRONMENTALISM

1. How would you describe the environment?

____
____
____
____

2   According to your knowledge, the term biodiversity can best be described as:
( ) Diversity of living beings.
(   ) Association of several hierarchical components: ecosystem, community, species, populations and genes in a defined area.
( ) The totality of genes, species and ecosystems in a region.

3. What is the best way to describe the changes that could occur to this biodiversity?
( ) Biodiversity can be seen as static.
( ) Biodiversity is immersed in a dynamic of changes that occur all the time.
(   ) It can be said without a doubt that biodiversity is dynamic, being distributed equally on Earth, in general terms and over time.

4   Generally, this biodiversity is altered by environmental impacts. These, in turn, can be better defined as:
( ) Negative changes in the environment caused by human actions.
( ) Major changes in the environment caused by human actions.
( ) Negative and positive, large or small changes in the environment caused by human actions.

5. Another factor that can directly affect biodiversity in various locations is global warming, which can currently be better defined as:
( ) A natural process that occurs on planet Earth and is a direct result of the rise in temperature caused by different factors.
( ) A process caused directly by greenhouse gas emissions.
( ) A natural process that occurs on planet Earth and is accelerated by greenhouse gas emissions.

ENVIRONMENTALISM AT UFPI

1)   In your opinion, how concerned is UFPI as an institution about maintaining and preserving the environment?
( ) Good, because UFPI has a set of measures aimed at this approach that have produced good results.
( ) Fair, because the measures taken by UFPI are not exactly sufficient, but they do help with this approach.
(   ) Bad, because none of the measures are effective or there are no measures at UFPI for such an approach.

2 Should UFPI adopt projects aimed at preserving the environment and disseminating environmental knowledge?
( ) Yes   ( ) No   ( ) Maybe

1 In your opinion, what measures should UFPI adopt in order to maintain and preserve the environment?
____
____
____
____

4 What about undergraduates as a group, what is the best alternative they should look for in order to disseminate environmental knowledge?

(   ) Organising events such as environmental week and/or exhibitions posters with messages about this approach.

( ) Dissemination of science fairs in the block.

( ) Symposiums, mini-courses, lectures, regular spaces for developing work, developing projects in this approach, etc.

5L In your opinion, what are the best benefits generated by UFPI's implementation of a set of measures aimed at maintaining and preserving the environment?

( ) Raising awareness among undergraduates, teachers, technicians and others.

( ) The consequent preservation of the environment and the sustainable use of spaces at UFPI, in terms of buildings.

( ) Maintaining green areas and reducing environmental damage at UFPI.

SELECTIVE COLLECTION

1. What is selective collection?

____

____

____

The prior separation of the various types of rubbish and waste begins with the waste containers, which have different colours for each type of rubbish. With regard to the colours of waste containers, the following can be said:

( ) The colour green is for plant materials, blue for plastics and yellow for paper / cardboard.

(   ) The colour brown is for organic waste, green is for glass and white for hospital/healthcare waste.

(   ) The colours purple, orange and grey are not used in containers of waste.

2. Among the methods used to reuse or process rubbish, which of the following is the most environmentally friendly?

(   ) Landfill, where rubbish is deposited under layers of earth with collection or incineration of methane gas.

( ) The incineration of rubbish, which drastically reduces the space it takes up in the environment.

( ) Recycling and composting to reuse rubbish.

3. The great importance of selective collection can be better attributed to:

( ) Reducing costs for the public purse, which consequently doesn't have to spend so much looking for new spaces to build landfill sites or rubbish dumps.

( ) Benefit to the environment, as separating rubbish directly enables the recycling process, as well as reducing the volume of rubbish for final disposal.

( ) Facilitating the work of recyclers and waste collectors.

4. The factors that can hinder the success of selective collection are best represented by:

( ) The lack of structures or places to process rubbish separately (recycling and composting).

(   ) Only a lack of interest on the part of the government itself in drawing up plans for the success of selective collection.

( ) Lack of public awareness and government incentives.

GRADUATION

1. What did the degree course in biological sciences bring you in terms of environmental knowledge?
( ) It allowed me to learn and improve on what I already knew.
( ) It gave me completely new information, knowledge and content. () It didn't give me any new environmental knowledge.

2. In its conception, the subjects on offer, which deal with themes linked to environmental knowledge, are:
(   ) Sufficient, but they need to improve their approaches in this regard environmental.
(   ) Insufficient, as they do not address this environmental aspect at all understandable or direct form.
( ) I don't have an elaborate concept.

3. In your opinion, who should demand that the various aspects of environmental knowledge be addressed?
( ) By the teachers and students themselves.
( ) By the course coordinator.
( ) For everyone.

4. What environmental knowledge did you acquire during your degree course in biological sciences?

____

____

____

____

5. In your opinion, your level of performance in the search for environmental knowledge is:
( ) Good, because I try to find out about it or keep up to date with it.
( ) Fair, because it's a topic that catches my attention, but doesn't hold it that much.
( ) Bad, because I'm not usually interested in these approaches.

Buy your books fast and straightforward online - at one of world's fastest growing online book stores! Environmentally sound due to Print-on-Demand technologies.

Buy your books online at
www.morebooks.shop

Kaufen Sie Ihre Bücher schnell und unkompliziert online – auf einer der am schnellsten wachsenden Buchhandelsplattformen weltweit! Dank Print-On-Demand umwelt- und ressourcenschonend produziert.

Bücher schneller online kaufen
www.morebooks.shop

Printed by Books on Demand GmbH, Norderstedt / Germany